ENERGY SECTOR STANDARD
OF THE PEOPLE' S REPUBLIC OF CHINA

中华人民共和国能源行业标准

Technical Specification for Construction and Installation of Onshore Wind Power Projects

陆上风电场工程施工安装技术规程

NB/T 10087-2018

Chief Development Department: China Renewable Energy Engineering Institute

Approval Department: National Energy Administration of the People's Republic of China

Implementation Date: March 1, 2019

China Water & Power Press
中国水利水电出版社
Beijing 2024

图书在版编目（CIP）数据

陆上风电场工程施工安装技术规程 : NB/T 10087-2018 = Technical Specification for Construction and Installation of Onshore Wind Power Projects（NB/T 10087-2018） : 英文 / 国家能源局发布. -- 北京 : 中国水利水电出版社, 2024. 3. -- ISBN 978-7-5226-2671-0

Ⅰ. TM614-65

中国国家版本馆CIP数据核字第2024RJ1450号

ENERGY SECTOR STANDARD
OF THE PEOPLE'S REPUBLIC OF CHINA
中华人民共和国能源行业标准

Technical Specification for Construction and Installation of Onshore Wind Power Projects

陆上风电场工程施工安装技术规程

NB/T 10087-2018

（英文版）

Issued by National Energy Administration of the People's Republic of China
国家能源局　发布
Translation organized by China Renewable Energy Engineering Institute
水电水利规划设计总院　组织翻译
Published by China Water & Power Press
中国水利水电出版社　出版发行
Tel: (+ 86 10) 68545888　68545874
sales@mwr.gov.cn
Account name: China Water & Power Press
Address: No.1, Yuyuantan Nanlu, Haidian District, Beijing 100038, China
http: //www.waterpub.com.cn
中国水利水电出版社微机排版中心　排版
北京中献拓方科技发展有限公司　印刷
184mm×260mm　16 开本　4.75 印张　150 千字
2024 年 3 月第 1 版　2024 年 3 月第 1 次印刷
Price（定价）: ¥ 760.00

Introduction

This English version is one of China's energy sector standard series in English. Its translation was organized by China Renewable Energy Engineering Institute authorized by National Energy Administration of the People's Republic of China in compliance with relevant procedures and stipulations. This English version was issued by National Energy Administration of the People's Republic of China in Announcement [2023] No. 1 dated February 6, 2023.

This version was translated from the Chinese Standard NB/T 10087-2018, *Technical Specification for Construction and Installation of Onshore Wind Power Projects*, published by China Water & Power Press. In the event of any discrepancy in the implementation, the Chinese version shall prevail.

Many thanks go to the staff from the relevant standard development organizations and those who have provided generous assistance in the translation and review process.

For further improvement of the English version, any comments and suggestions are welcome and should be addressed to:

China Renewable Energy Engineering Institute
No. 2 Beixiaojie, Liupukang, Xicheng District, Beijing 100120, China
Website: www.creei.cn

Translating organizations:

POWERCHINA Northwest Engineering Corporation Limited

Northwest Water & Hydropower Engineering Co., Ltd.

Translating staff:

QIAO Peng LI Kejia BAI Xueyuan ZHOU Danshun

WANG Yindong

Review panel members:

GUO Jie POWERCHINA Beijing Engineering Corporation Limited

HOU Yujing China Institute of Water Resources and Hydropower Research

ZHANG Ming Tsinghua University

YOU Yang	China Society for Hydropower Engineering
LI Qian	POWERCHINA Zhongnan Engineering Corporation Limited
LI Yu	POWERCHINA Huadong Engineering Corporation Limited
WANG Yawen	POWERCHINA Chengdu Engineering Corporation Limited

National Energy Administration of the People's Republic of China

翻译出版说明

本译本为国家能源局委托水电水利规划设计总院按照有关程序和规定，统一组织翻译的能源行业标准英文版系列译本之一。2023年2月6日，国家能源局以2023年第1号公告予以公布。

本译本是根据中国水利水电出版社出版的《陆上风电场工程施工安装技术规程》NB/T10087—2018翻译的，著作权归国家能源局所有。在使用过程中，如出现异议，以中文版为准。

本译本在翻译和审核过程中，本标准编制单位及编制组有关成员给予了积极协助。

为不断提高本译本的质量，欢迎使用者提出意见和建议，并反馈给水电水利规划设计总院。

地址：北京市西城区六铺炕北小街2号
邮编：100120
网址：www.creei.cn

本译本翻译单位：中国电建集团西北勘测设计研究院有限公司
西北水利水电工程有限责任公司

本译本翻译人员：乔　鹏　李可佳　白雪源　周丹顺
王银东

本译本审核人员：

郭　洁　中国电建集团北京勘测设计研究院有限公司
侯瑜京　中国水利水电科学研究院
张　明　清华大学
由　洋　中国水力发电工程学会
李　倩　中国电建集团中南勘测设计研究院有限公司
李　瑜　中国电建集团华东勘测设计研究院有限公司
王雅雯　中国电建集团成都勘测设计研究院有限公司

国家能源局

Announcement of National Energy Administration of the People's Republic of China [2018] No. 12

According to the requirements of Document GNJKJ [2009] No. 52 "Notice on Releasing the Energy Sector Standardization Administration Regulations (*tentative*) and detailed implementation rules issued by National Energy Administration of the People's Republic of China", 204 sector standards such as *Coal Mine Air-Cooling Adjustable-Speed Magnetic Coupling*, including 54 energy standards (NB), 8 petrochemistry standards (NB/SH) and 142 petroleum standards (SY), are issued by National Energy Administration of the People's Republic of China after due review and approval.

Attachment: Directory of Sector Standards

National Energy Administration of the People's Republic of China

October 29, 2018

Attachment:

Directory of Sector Standards

Serial number	Standard No.	Title	Replaced standard No.	Adopted international standard No.	Approval date	Implementation date
…						
42	NB/T 10087-2018	Technical Specification for Construction and Installation of Onshore Wind Power Projects			2018-10-29	2019-03-01
…						

Foreword

According to the requirements of Document GNKJ [2014] No. 298 issued by National Energy Administration of the People's Republic of China, "Notice on Releasing the Development and Revision Plan of the First Batch of Energy Sector Standards in 2014", and after extensive investigation and research, summarization of practical experience, and wide solicitation of opinions, the drafting group has prepared this specification.

The main technical contents of this specification include: construction preparation, civil works construction, wind turbine installation, electrical equipment installation, debugging and commissioning, quality inspection and control, and occupational health, safety and environmental protection.

National Energy Administration of the People's Republic of China is in charge of the administration of this specification. China Renewable Energy Engineering Institute has proposed this specification and is responsible for its routine management. The Sub-committee on Construction and Installation of Wind Power Project of Energy Sector Standardization Technical Committee on Wind Power is responsible for the explanation of specific technical contents. Comments and suggestions in the implementation of this specification should be addressed to:

China Renewable Energy Engineering Institute
No. 2 Beixiaojie, Liupukang, Xicheng District, Beijing 100120, China

Chief development organizations:

POWERCHINA Northwest Engineering Corporation Limited

Northwest Water & Hydropower Engineering Co., Ltd.

Participating development organization:

Xinjiang Goldwind Science & Technology Co., Ltd.

Chief drafting staff:

ZHOU Caigui	WANG Yuwu	BAI Xueyuan	LUO Yinhe
CAO Bizheng	JIN Gang	SHEN Kuanyu	HUANG Wenxiang
HAN Rui	WU Jiang	SUN Wei	LI Zhongxiang
ZHANG Wei	XU Haijun	LI Zirong	WANG Xinquan
WANG Shuanglin	WANG Yindong	WANG Guanping	WU Hongkui
ZHANG Miaofang	JIA Jiuming	JI Zhenhua	MING Jing

Review panel members:

CHANG Zuowei	WANG Yi	FAN Xiaoping	YU Jialin
ZHANG Jie	ZHU Kainian	GAO Pengfei	CHEN Guibin
LU Tianyu	TANG Qunyi	WU Chengzhi	QIU Changkai
CHEN Kangdong	JI Hao	ZHANG Aishun	ZHANG Liangjun
ZHAO Min	SHUAI Zhengfeng	LI Xiang	LI Shisheng

Contents

1 General Provisions

1.0.1 This specification is formulated with a view to standardizing the construction and installation works and ensuring the construction and installation quality and safety for onshore wind power projects.

1.0.2 This specification is applicable to the construction and installation works of onshore wind power construction, renovation or extension projects.

1.0.3 In addition to this specification, the construction and installation works of onshore wind power projects shall comply with other current relevant standards of China.

2 Terms

2.0.1 wind speed in construction work

10 min average wind speed at hub height of the wind turbine

2.0.2 maximum wind speed for erection

upper-limit wind speed for wind turbine lifting and installation

2.0.3 foundation connector

member embedded in the foundation to connect the foundation and the tower, which can be categorized as a foundation ring, bolted cage, or other types

3 Basic Requirements

3.0.1 The construction and installation works of onshore wind power projects shall include the construction preparation, civil works construction, wind turbine installation, electrical equipment installation, debugging and commissioning.

3.0.2 The construction and installation contractors of onshore wind power projects shall have pertinent qualifications, and the practitioners shall have relevant qualifications.

3.0.3 A sound management system for quality control, occupational health and safety, and environmental protection shall be established and implemented in accordance with the management objectives.

3.0.4 The construction and installation works shall be implemented by the contractor in accordance with the technical documentation reviewed and signed by the supervisor.

3.0.5 A use permit shall be obtained when new technology, new process, new equipment or new material is applied in the construction, in which case, a special work statement shall be formulated and implemented after approval.

3.0.6 The construction and installation data of the entire process shall be collected, sorted out and compiled in accordance with the current sector standard NB/T 31118, *Archives Acceptance Specification for Wind Power Projects* to ensure their authenticity, validity and completeness. The data shall be timely handed over to the owner after the project is completed.

4 Construction Preparation

4.1 Technical Preparation

4.1.1 The following data shall be collected before the construction and installation is initiated:

1 Hydrological and meteorological data of the project area.

2 Engineering geological data of building (structure) sites.

3 Survey data of buildings (structures) and neighbouring underground pipelines, underground buildings (structures), abandoned underground buildings (structures), tombs, caves, etc.

4 Technical specification data and installation requirements for wind farm equipment and their accessories.

5 Construction design documents.

4.1.2 Before the project starts, the control network provided by the owner shall be checked by the contractor in accordance with the contract requirements.

4.1.3 Prior to construction and installation, the supervisor shall organize meetings for design disclosure and drawings review. The contractor shall, in accordance with the design documents and the site-specific situation, compile the construction planning, method statement or work instruction, and submit them to the supervisor for review and approval.

4.1.4 The contractor shall identify the dangerous and harmful factors according to the project scale and structure grade of the wind power project. Before the construction of the works and sections with considerable risks, a special work statement shall be prepared and reported to the supervisor for approval and record. As for important temporary facilities, important construction procedures, special activities, and dangerous operations, special safety protection and technical measure reports shall be prepared and submitted to the supervisor for approval; as for the foundation pit blasting excavation, tower and wind turbine transportation via mountain roads, tower and wind turbine handling and erection, and other dangerous operations, safety permits shall be obtained before construction starts.

4.1.5 The construction and configuration of fire protection facilities shall comply with the current national standard GB 50720, *Technical Code for Fire Safety of Construction Site*.

4.1.6 The contractor shall formulate an environmental protection and civilized construction plan for the construction period in accordance with the environmental and ecological conditions of the project area as well as the requirements of local laws and regulations.

4.1.7 The contractor shall organize safety training and technical disclosure for the practitioners. Special post personnel and special equipment operators shall be certified to work.

4.1.8 The detection and testing devices, instruments and tools used in the construction and installation shall be checked and calibrated according to relevant regulations and used within the valid period.

4.1.9 For construction activities in winter and rainy seasons, measures such as anti-freezing, anti-slip and flood protection shall be taken, and method statements for construction safety and quality assurance shall be formulated.

4.2 On-Site Preparation

4.2.1 The contractor shall set up a project management department staffed with managerial and technical personnel. The key managerial and technical personnel shall have corresponding capabilities and qualifications.

4.2.2 The project management department shall be equipped with office supplies, measuring instruments, inspection instruments, transportation vehicles, and living facilities.

4.2.3 Temporary facilities for construction shall be reasonably arranged according to the site construction needs.

4.2.4 The roads, power supply, water supply, communications and site grading shall be completed before construction commencement.

4.2.5 The performance of construction machinery and equipment shall meet the requirements for construction and installation. The use of construction machinery and equipment shall comply with the requirements of corresponding mechanical operating procedures and shall pass the inspections.

4.2.6 The inspection of lifting appliances shall comply with the current national standard GB/T 5905, *Cranes —Test Code and Procedures*.

4.2.7 The temporary power use at the construction site shall comply with the current national standard GB 50194, *Code for Safety of Power Supply and Consumption for Construction Site*.

4.2.8 The construction site shall have drainage measures and meet the flood control requirements.

4.2.9 Lightning protection shall be provided for important temporary facilities such as inflammable and explosive materials warehouses, generator rooms, power distribution rooms, boiler rooms, temporary oil depots, as well as high-rise temporary facilities such as concrete batching plant, cement and fly ash silos. The lightning protection devices shall comply with the current national standard GB 50057, *Code for Design Protection of Structures Against Lightning*.

4.2.10 Inflammable and explosive materials warehouses, generator rooms, power distribution rooms, boiler rooms, living houses, kitchens and other temporary buildings shall not be constructed of flammable materials, and shall be provided with necessary protective measures.

4.2.11 Appropriate warning and safety signs shall be set at the construction site.

4.3 Equipment Transportation and Storage

4.3.1 Before transportation of the large or heavy equipment, special technical plans for loading and unloading the large or heavy equipment shall be prepared. Designated personnel shall be assigned to instruct the operators and supervise the safety during the loading and unloading operations.

4.3.2 Before transportation of the large or heavy equipment, the equipment shall be properly protected and fastened according to the requirements against deformation, impact, corrosion, and damping.

4.3.3 Windproof, rainproof, dustproof and anti-vibration measures shall be taken during equipment transportation and storage at site. The supervisor, wind turbine installation contractor, and equipment manufacturer shall jointly participate in the unpacking and acceptance of the equipment delivered to the site.

4.3.4 Transportation of the large or heavy equipment shall meet the following requirements:

1 Before transportation, the vehicles shall be subjected to the carrying capacity check, and the stability check under specific road conditions during transportation along the designated route.

2 During loading and unloading, the equipment shall be lifted up and down steadily and orderly. During transportation, loading, unloading and storage, the vulnerable parts, bolts and matching surfaces of the equipment shall be particularly protected.

3 During transportation, the center of gravity of the equipment shall coincide with the central axis of the vehicle. The equipment shall be securely fastened and labeled with warning signs.

4 During transportation, the vehicles shall run at a controlled speed and turn slowly to ensure stable transportation.

5 Anti-deformation measures shall be taken for towers in transportation and storage, and protective measures shall be taken for coatings and joint surfaces. Equipment in the nacelle shall be protected from water or corrosion.

6 The transportation and storage of the main transformer shall comply with the current national standard GB 50148, *Code for Construction and Acceptance of Power Transformers Oil Reactor and Mutual Inductor*.

4.3.5 The laydown yard for large or heavy equipment shall be arranged according to the topographical and geological conditions, layout of wind turbines, and transport conditions. The yard shall meet the equipment handling and safety requirements.

5 Civil Works Construction

5.1 General Requirements

5.1.1 The raw materials, ready-mixed concrete, prefabricated components, etc. for the project shall have the certificates of conformity before leaving the factory, and shall be re-inspected and confirmed to be qualified before use in accordance with the current national standard GB 50204, *Code for Quality Acceptance of Concrete Structure Construction*.

5.1.2 Mix proportion design and trial batching shall be performed before concrete construction. The mix ratio shall meet the specified properties and construction requirements.

5.1.3 The capacities of concrete production, transportation and pouring equipment shall meet the construction intensity requirements.

5.1.4 At each construction stage, the works inspection and acceptance shall be carried out in accordance with the three-level inspection system, i.e. self-inspection, mutual inspection and handover inspection, and the key processes and concealed works shall be inspected and recorded.

5.2 Construction of Road and Crane Pad

5.2.1 Before construction of the roads and crane pads, the original ground shall be surveyed and set out. The setting-out shall meet the following requirements:

1 Before construction of the road subgrade, set out the centerline of the whole route, set up the main control stakes, and resurvey the original ground; set up edge pegs for the subgrade boundaries, embankment toe, cutting top, borrow area, and spoil area.

2 Before construction of the crane pad, conduct setting-out to determine the construction site boundary and edge peg locations, and set up markers such as the sidelines and centerlines of the structures.

5.2.2 The land areas for the roads, crane pads and fill locations shall be stripped of topsoil, and the topsoil cleared shall be piled up in a concentrated manner. The areas for roads and crane pads shall be compacted after site grading according to the compactness requirements.

5.2.3 The width, flatness, slope and subgrade bearing capacity of the roads in the construction site shall meet the requirements for transportation of large or heavy equipment, and safety protection measures shall be taken when the equipment are placed on slopes.

5.2.4 The main technical parameters of roads in the wind farm shall be in accordance with the current sector standard NB/T 31113, *Code for Construction Organization Design of Onshore Wind Power Projects*. Road construction shall meet the following requirements:

1 Before road construction, detailed investigation shall be conducted on the geology, hydrology, obstacles, cultural relics and various pipelines within the construction site, and records shall be made.

2 During road construction, cut and fill shall be well balanced and construction procedures shall be reasonably arranged, to reduce disposal and minimize impact on the environment.

3 The road subgrade and pavement construction shall comply with the current sector standards JTG F10, *Technical Specification for Construction of Highway Subgrades*; and JTG/T F20, *Technical Guidelines for Construction of Highway Roadbases*.

4 In dangerous sections such as high embankments, steep slopes, sharp bends, and mountain roads, safety facilities such as speed limits, alerts, warnings, direction signs and protective fences shall be set up on both sides of the road during the construction period.

5.2.5 In addition to the requirements of the design and the wind turbine manufacturer, the crane pad shall also meet the following requirements:

1 The size of the crane pad shall meet the requirements for installation and operation of wind turbines, and the crane pad shall be smoothly connected with the road on site.

2 The backfilled ground shall be filled and compacted by layers, and the compactness shall meet the design requirements.

3 The excavated ground should be stripped of topsoil down to the undisturbed level and graded as the crane pad, and the bearing capacity of the ground shall meet the design requirements.

4 The crane pad shall have a slope of 1 % to 2 % to facilitate drainage. When there is a cut slope, a side drainage ditch shall be excavated at the slope toe, and the inside of the ditch shall be sealed with cement mortar. When the catchment area at the cut slope side is large, an intercepting ditch shall be set at the upper edge of the cut slope. The intercepting ditch should not be less than 5 m away from the edge of the cut slope, and the inside of the ditch shall be sealed with cement mortar.

5.3 Foundation Construction for Wind Turbine and Prefabricated Substation

5.3.1 The foundation excavation shall meet the following requirements:

1 Before excavation, a construction plan shall be developed according to the geological conditions, excavation depth, construction method, loads on ground, design slope ratio and other data. When the excavation depth exceeds 3 m, a special safety plan shall be developed.

2 Earthwork excavation shall be conducted in layers from top to bottom to avoid local excessive excavation and overspeed unloading which may lead to soil mass destabilization. When the foundation pit is excavated mechanically, a 300 mm thick base layer shall be reserved for artificial cutting to avoid disturbing the foundation surface.

3 Mechanical means are preferred for rockwork excavation. When blasting operation is necessary, the controlled blasting or pre-split blasting should be adopted. Blasting operations shall meet the requirements of the current national standard GB 50201, *Code for Construction and Acceptance of Earthwork and Blasting Engineering*.

4 During earth-rock work excavation, the earth, rock, materials and equipment should not be stacked nearby the foundation pit. If such stacking is inevitable, the stacked materials and the slope of the foundation pit shall be stable, and safety measures shall be taken as necessary.

5 After the foundation pit is excavated to the design elevation, geological logging shall be made, and the pit ground shall be jointly inspected by the owner, designer, surveyor and supervisor.

6 After the pit ground is inspected and accepted, the foundation cushion layer shall be constructed timely; if it is not cushioned over 48 h, the pit ground shall be re-inspected. When the foundation pit is soaked, frozen to heaving, or disturbed, the affected soil in the pit shall be removed, and the pit shall be treated, re-inspected and accepted according to the design requirements.

5.3.2 In addition to the current standards of China GB 50202, S*tandard for Acceptance of Construction Quality of Building Foundation*; and JGJ 79, *Technical Code for Ground Treatment of Buildings*, the ground treatment shall also meet the following requirements:

1 For ground treatment by compaction and replacement, the excavation depth, replacement materials testing, construction methodology, layered placement thickness, compaction passes, moisture content, compactness, and drainage measures shall be well controlled or selected according to the design requirements.

2 For ground treatment with compaction piles, the hole position, depth and diameter, filling materials, construction parameters, etc. shall be well controlled according to the design requirements, and the construction records shall be made truthfully.

3 The precast concrete pile foundation shall comply with the current sector standard JGJ 94, *Technical Code for Building Pile Foundations,* and meet the design requirements. After construction, anti-corrosion measures shall be taken for the exposed steel rings. The pile heads shall be firmly connected with the foundation of the wind turbine pile cap, and a supporting plate shall be set inside the pile head.

4 The bored cast-in-place pile foundation shall comply with the current sector standard JGJ 94, *Technical Code for Building Pile Foundations,* and meet the design requirements. The length of the pile head and the longitudinal main rebar embedded in the pile cap shall meet the design requirements; the length of the pile head embedded in the cap shall not be less than 50 mm, and the anchorage length of the longitudinal main rebar embedded in the cap shall not be less than 35 times the diameter of the longitudinal main rebar. For uplift piles, the anchorage length of the longitudinal main rebars at the pile top shall be in accordance with the current national standard GB 50010, *Code for Design of Concrete Structures*.

5 The rock-anchored foundation shall comply with the current national standard GB 50086, *Technical Code for Engineering of Ground Anchorages and Shotcrete Support,* and meet the design requirements. During construction, the anchor location, borehole diameter, depth and angle, insertion length, grout mix ratio, grouting pressure, grout take, anchor stress, etc., shall be checked.

5.3.3 The construction related to embedded parts and buried pipes in foundation shall meet the following requirements:

1 The embedded parts shall be set in the right position and direction and fixed securely. Surface embedded parts shall be provided with positioning measures. If the embedded parts conflict with the rebars, the spacing of rebars may be adjusted appropriately, and the rebars shall not be reduced or cut off.

2 The orientation of foundation connector supports shall meet the design requirements.

3 Cable ducts and drainage pipes shall be buried according to the design requirements, fixed firmly, and well protected to prevent debris from entering.

4 Before concrete pouring, the centerline, elevation, orientation and fixing of the embedded parts, embedded pipes and reserved holes shall be checked and accepted.

5.3.4 Preparation before installation of the foundation connectors shall meet the following requirements:

1 For a ring-type foundation connector, check the anticorrosive coating, reserved holes, and waterproof conditions of the foundation ring, and repair them if any damage.

2 For a bolted cage-type connector or other types of foundation connectors, check the components quantity and their appearance. Unqualified products are prohibited from use.

3 Before installation of the foundation connectors, check whether the erection site, hoisting machinery, lifting slings, and operators can meet the safety requirements, and prohibit any rule-breaking operations.

5.3.5 The foundation connector installation shall meet the following requirements:

1 Before installation, the strength of the cushion concrete shall reach more than 70 % of the design strength.

2 The installation elevation shall meet the requirements in the design drawings.

3 After putting in place, check the center of the foundation connector to ensure that its projection center coincide with the foundation center.

4 After installation in place, correct the verticality of the support. The foundation connector and the main support shall be vertical.

5 When welding the foundation connector with the support, it is forbidden to disengage the crane. The firmness of the foundation connectors and supports shall meet the design requirements.

5.3.6 The foundation connector leveling shall meet the following requirements:

1 After the foundation connector is installed in place, it should be leveled during a period of no wind or of breeze. A record shall be made during leveling such as weather and temperature conditions.

2 The leveling error of the foundation connector shall meet the technical requirements of both the designer and the manufacturer.

3 After the foundation connector leveling is done, its levelness shall be re-checked before, during and after concrete pouring, and duly recorded.

5.3.7 The levelness control for the foundation connector shall meet the following requirements:

1 The fixing frame of the foundation connector shall be a self-sustained system, independent of the rebars and the formwork support system.

2 During installation, the rebars shall not contact the foundation connector. Rebars laterally running through the foundation connector shall be suspended in the holes to avoid contact with the foundation connector. The fabrication and installation of rebars on the foundation connector are prohibited.

3 The foundation concrete shall be poured up evenly to prevent the foundation connector from deflection due to unbalanced squeeze caused by the pour height difference.

4 The exposed parts of the foundation connector shall be covered and protected during concrete pouring, and at the same time, any objects or facilities shall avoid contacting the foundation connector to prevent its deflection.

5 When the concrete is poured to the adjusting bolts of the foundation connector, measure the levelness to make sure that the levelness error is within the permissible range.

5.3.8 Rebar fabrication and installation shall meet the following requirements:

1 Rebar connection shall comply with the current sector standards JGJ 107, *Technical Specification for Mechanical Splicing of Steel Reinforcing Bars*; JG/T 163, *Couplers for Rebar Mechanical Splicing*; and JGJ 18, *Specification for Welding and Acceptance of Reinforcing Steel Bars*.

2 Rebars shall be fixed with reliable supports and spacers, pads, etc. Rebar and stone shall not be used as pads. The upper reinforcement of the wind turbine foundation, when it is not clearly defined in design, may be fixed by a steel bracket which shall ensure the stability of the upper reinforcement.

3 The allowable position deviation of the reinforcement and the thickness of the concrete cover shall meet the design requirements.

4 When preparing the reinforcement, the laying of cable ducts, drainage pipes and lightning protection earthing shall be done in time.

5.3.9 The foundation formwork fabrication and installation shall meet the following requirements:

1 The formwork construction shall comply with the current sector standard JGJ 162, *Technical Code for Safety of Forms in Construction*. Structural members such as steel pipes and couplers shall comply with the current sector standard JGJ 130, *Technical Code for Safety of Steel Tubular Scaffold with Couplers in Construction*, and may only be used after acceptance.

2 The formwork and its support system shall be selected and designed according to the actual situations. The foundation for the wind turbine should adopt the composite steel formwork.

3 The formwork and its support system shall have sufficient bearing capacity, stiffness and stability, and be capable of bearing the poured concrete weight, lateral pressure and construction loads.

4 The design and construction of the formwork shall be simple, easy to assemble and dismantle, and meet the technical requirements for installation, concrete pouring and curing.

5 Joints of the formwork shall be tight and close-fitting, and measures shall be taken to prevent slurry leaking.

6 The composite steel-formwork shall comply with the current national standard GB/T 50214, *Technical Code for Composite Steel-Form*. The steel formwork and its bearing pieces shall be properly stored and maintained. Steel forms with severe corrosion, deformation and cracks shall not be used. Other types of formwork shall comply with the current sector standards JGJ/T 74, *Technical Standard for Large-Area Formwork of Building Construction*; JGJ 96, *Technical Specification for Plywood Form with Steel Frame*; and JG/T 156, *Plybamboo Form*.

7 The form surface shall be coated with release agent. The release agent shall not contaminate the reinforcement or affect the appearance of exposed structures.

8 The installation deviation of the formwork shall meet the design

requirements.

5.3.10 The foundation concrete construction shall meet the following requirements:

1 Before concrete construction, a special work statement shall be formulated, technical issues shall be clarified, and transport, material and power supply shall be ensured. Furthermore, according to the slump loss of concrete, the initial setting time, as well as the mixing and transportation capacity, the pouring sequence and distribution points shall be determined to prevent construction cold joints.

2 The concrete pumping construction shall meet the design requirements and comply with the current sector standard JGJ/T 10, *Technical Specification for Construction of Concrete Pumping*.

3 The formwork and embedded parts shall be inspected and accepted before concrete pouring, and be checked and maintained by dedicated personnel during pouring.

4 Concrete shall be poured in layers, continuously, compacted by tamping, one-time finished, and leave no construction joints.

5 When the wind turbine foundation is constructed of mass concrete, it shall meet the design requirements and comply with the current national standard GB 50496, *Standard for Construction of Mass Concrete*.

6 Concrete sampling and specimen preparation shall comply with the current national standard GB 50204, *Code for Quality Acceptance of Concrete Structure Construction*.

7 Temperature measurement points shall be arranged according to the design requirements. Temperature measurement shall comply with the current national standard GB/T 51028, *Technical Code for Temperature Measurement and Control of Mass Concrete*, and records shall be made.

5.3.11 Only when the concrete strength can ensure that its surface and edges are not damaged, can the side form be removed. To dismantle the load-bearing form and its support, the concrete strength shall meet the relevant requirements in GB 50666, *Code for Construction of Concrete Structures*. To remove the form in a reserved hole, no cave-in or cracking of the concrete surface shall be ensured, and dismantling the form shall avoid causing vibration or damage to the hole wall.

5.3.12 Concrete curing shall meet the following requirements:

1 The curing time for mass concrete of the wind turbine foundation shall not be less than 14 d.

2 Cover and moisturize the concrete within 12 h after pouring. When the average temperature in a day is below 5 °C, no watering is allowed. Watering times shall be so determined to keep the concrete in a wet state. The water for concrete curing shall be the same as that for concrete mixing.

3 Before the concrete strength reaches 1.2 MPa, no stepping or loading on it is allowed.

5.3.13 The anti-corrosion construction for foundation shall meet the design requirements and comply with the current national standards GB 50212, *Code for Construction of Building Anticorrosive Engineering*; and GB 50224, *Code for Acceptance of Construction Quality of Anticorrosive Engineering of Buildings*. The base surface for anti-corrosion coating shall be compact, level, and free of pollution or defects. The finished anti-corrosion coating shall be protected according to the design requirements.

5.3.14 The anti-corrosion coating construction shall meet the following requirements:

1 When applying any new types of coating, the coating materials shall be subjected to property test and application test to meet the relevant quality requirements.

2 The base shall be treated according to the design requirements. The base concrete surface of the dry-based coating shall be dry, and there shall be no obvious water accumulation on the base concrete surface for the moisture-curing coating.

3 The coating shall be prepared in accordance with the prescribed mix ratio and the mix procedure under the protection of facilities against sunshine, rainfall, wind, and sand.

4 The ambient temperature during coating construction shall meet the requirements of the product manual.

5 The construction site shall have fire control and poison protection measures, and construction personnel shall be provided with protective equipment.

6 The coating construction shall not proceed until the coating surface dries.

7 The coating material shall be stirred uniformly, the thickness of the coating shall be consistent, and there shall be no coat missing, wrinkles, bubbles, and film break.

5.3.15 The construction of anti-corrosive coiled material shall meet the following requirements:

1 The coiled material shall be bonded tightly layer by layer, and free of bubbles, cracks or delamination.

2 All corners shall be rounded, and measures shall be taken to protect the coiled material.

3 When bonding the coiled material, the overlap width shall meet the design requirements. The overlaps between two adjacent sheets or between the upper and lower layers shall be staggered, and not be bonded perpendicular to each other.

5.3.16 In addition to the technical requirements of the design and the manufacturer, the waterproof seal between the foundation ring and the concrete foundation shall meet the following requirements:

1 The working face shall be cleared during construction to ensure that the base surface is free from looseness, dust, water, burrs, contamination and defects.

2 The ambient temperature during construction shall meet the requirements of the product manual.

5.3.17 After foundation construction, only when the anti-corrosion treatment is completed and accepted in accordance with the design requirements, can the foundation backfill start. The backfill construction shall meet the following requirements:

1 The backfill construction shall comply with the current national standard GB 50202, *Standard for Acceptance of Construction Quality of Building Foundation*.

2 Before backfilling, it shall be ensured that there are no wastes, tree roots or debris, accumulated water, mud, ice, etc. in the foundation pit.

3 Backfill shall start from the lowest place and be roller-compacted in layers, and the compactness shall meet the design requirements.

4 The moisture content of backfill soil shall be tested in the rainy season. Backfill soil construction is not allowed in rainy and snowy days, and frozen soil shall not be used for backfilling. After the completion of backfilling, temporary drainage measures for the site shall be taken.

5.3.18 Observation of the wind turbine foundation settlement shall meet the following requirements:

1 The first observation is made after the completion of foundation construction; the second observation is made when the foundation concrete strength reaches 70 % of the design strength; and the third and fourth observations are made before and after the wind turbine installation, respectively.

2 During construction, observations shall be increased in case of any special circumstances such as an earthquake of Magnitude 5 or above, a strong wind of Scale 8 or above, and flooding of the foundation .

3 For settlement observation, two independent sets of observations shall be performed each time. When the values of the two sets of observations do not exceed the prescribed value, the average of them shall be taken as the observation result. When the observation values exceed the prescribed value, the causes shall be analyzed.

4 In the settlement observation, complete observation records shall be made.

5.4 Civil Works for Collection Lines

5.4.1 The civil construction for collection lines shall minimize the land use and avoid or reduce its impact on the ground vegetation and the surrounding environment.

5.4.2 The civil construction for overhead lines shall meet the design requirements and comply with the current national standards GB 50173, *Code for Construction and Acceptance of 66 kV and Under Overhead Electric Power Transmission Line*; and GB 50233, *Code for Construction and Acceptance of 110 kV ~ 750 kV Overhead Transmission Line.*

5.4.3 The buried cable construction shall meet the design requirements and comply with the current national standard GB 50168, *Standard for Construction and Acceptance of Cable Line Electric Equipment Installation Engineering.*

5.4.4 The earthing connection construction shall meet the design requirements and comply with the current national standard GB 50169, *Code for Construction and Acceptance of Grounding Connection Electric Equipment Installation Engineering.*

5.5 Civil Works for Substation

5.5.1 Before construction is commenced, a surveying control network for the construction period shall be established, and the surveying and positioning shall meet the design requirements and comply with the current national standard GB 50026, *Code for Engineering Surveying.*

5.5.2 Building foundation construction shall meet the design requirements and comply with the current standards of China GB 50202, *Standard for Acceptance of Construction Quality of Building Foundation*; and JGJ 120, *Technical Specification for Retaining and Protection of Building Foundation Excavations.*

5.5.3 During excavation of foundation pits, ditches and trenches, slopes shall be formed according to design requirements and geological conditions, and measures shall be taken to ensure the stability of slopes, neighboring buildings, public facilities, temporary facilities, and the personnel safety. The pits, ditches and trenches after acceptance shall be closed immediately to reduce exposure and prevent water ingress, and the underground structure construction shall be carried out timely.

5.5.4 The underground waterproof construction shall meet the design requirements and comply with the current national standard GB 50208, *Code for Acceptance of Construction Quality of Underground Waterproof.*

5.5.5 The foundation corrosion protection shall be constructed in accordance with the design requirements and the current national standard GB 50212, *Code for Construction of Building Anticorrosive Engineering.*

5.5.6 The concrete construction shall meet the design requirements and comply with the current national standards GB 50666, *Code for Construction of Concrete Structures*; and GB 50204, *Code for Quality Acceptance of Concrete Structure Construction.*

5.5.7 The construction and quality acceptance of masonry structures shall meet the design requirements and comply with the current national standards GB 50924, *Code for Construction of Masonry Structures Engineering*; and GB 50203, *Code for Acceptance of Constructional Quality of Masonry Structures.*

5.5.8 The construction and acceptance of roof works shall meet the design requirements and comply with the current national standards GB 50345, *Technical Code for Roof Engineering*; and GB 50207, *Code for Acceptance of Construction Quality of Roof.*

5.5.9 The materials used in decoration shall be environmental friendly and safe. The materials prohibited by the state shall not be used. The indoor environmental pollution control for decoration works shall comply with the current national standards GB 50327, *Code for Construction of Decoration of Housings*; and GB 50325, *Code for Indoor Environmental Pollution Control of*

Civil Building Engineering. The burning characteristics of decorative materials shall comply with the current national standard GB 50222, *Code for Fire Prevention in Design of Interior Decoration of Buildings*.

5.5.10 The decoration construction shall meet the design requirements and comply with the current national standard GB 50327, *Code for Construction of Decoration of Housings*.

5.5.11 The external thermal insulation works on walls shall be constructed in accordance with the design requirements and the current sector standard JGJ 144, *Technical Specification for External Thermal Insulation on Walls*. The components shall be physically and chemically stable and shall meet the fire control requirements of the fire management agency.

5.5.12 After the trenches, concealed pipes, heat preservation, thermal insulation, sound proofing and other underground works are completed and inspected, and the concealed works have been accepted, the building works can be started. The building materials shall meet the design requirements, and the construction quality shall meet the design requirements and comply with the current national standard GB 50209, *Code for Acceptance of Construction Quality of Building Ground*.

5.5.13 The installation of water supply and drainage system shall meet the design requirements and comply with the current national standard GB 50242, *Code for Acceptance of Construction Quality of Water Supply Drainage and Heating Works*.

5.5.14 The construction and acceptance of the earthing devices shall comply with the design requirements and the current national standard GB 50169, *Code for Construction and Acceptance of Grounding Connection Electric Equipment Installation Engineering*. When the non-ferrous metal earth wire cannot be welded, it should be connected by bolting. The contact surface of the bolt connection shall meet the current national standard GB 50149, *Code for Construction and Acceptance of Busbar Installation of Electric Equipment Installation Engineering*.

5.5.15 The fabrication and installation of steel framework shall meet the design requirements and comply with the current standards of China GB 50205, *Code for Acceptance of Construction Quality of Steel Structures*; DL/T 646, *Manufacturing Technical Requirements for Steel Tubular Structures of Substation and Transmission Line*; DL 5190.1, *The Technical Specification for Electric Power Construction, Part 1: Civil Structure Work*; and DL/T 5210.1, *Specification for Construction Quality Acceptance and Evaluation of Electric*

Power Construct—Part 1: Civil Construction Engineering.

5.5.16 The indoor lighting system shall meet the design requirements and comply with the current sector standard DL/T 5161.17, *Specification for Construction Quality Checkout and Evaluation of Electric Equipment Installation—Part 17: Electrical Lighting Device.*

5.5.17 The fire protection system shall be designed, constructed and put into operation simultaneously with the main works. The system construction shall comply with the current national standards GB 50166, *Code for Installation and Acceptance of Fire Alarm System*; GB 50261, *Code for Installation and Commissioning of Sprinkler Systems*; and GB 50444, *Code for Acceptance and Inspection of Extinguisher Distribution in Buildings.*

5.5.18 The purchase, installation and acceptance of the ventilation and air-conditioning works shall meet the design requirements and comply with the current national standard GB 50243, *Code of Acceptance for Construction Quality of Ventilation and Air Conditioning Works.*

6 Wind Turbine Installation

6.1 General Requirements

6.1.1 Before the installation and construction of a wind turbine, the foundation, crane pad and construction road shall be inspected to ensure that the site conditions meet the requirements for large component transportation and wind turbine installation.

6.1.2 The equipment and materials for wind turbines shall be accompanied with quality certificates, and the high-strength bolts for connecting the tower shall also be subjected to quality re-inspection by a qualified testing organization.

6.1.3 The lifting machinery and slings shall be selected according to the weight of the wind turbine components, the center height of the hub, and the crane capacity parameters, and shall be inspected for safety before use. The inspection shall include the following:

1 Before using the lifting machinery, check their hydraulic system, slewing system, limit switch, anti-back-tilting device, winch and wire rope, and standing levelness.

2 Before using fiber slings, check their certificate, label, safe working load, and surface abrasion; before using rigid slings, check for bending deformation, cracking and damage phenomena.

6.1.4 The length of the tag line shall be calculated according to the equipment size and installation height; the ultimate tensile strength of the tag line shall meet the requirements for use.

6.1.5 The equipment lifting points shall be determined according to the technical requirements for the equipment. If there are no clear stipulations, they shall be determined based on calculation in terms of the equipment size, weight and gravity center.

6.1.6 Torque tools or high-strength bolt tensioning tools used for construction shall be within the valid period after verification and be calibrated on site before use.

6.1.7 High-strength bolts for structural connection shall be timely and symmetrically fastened several times to the specified torque, and anti-loosening signs shall be marked. The installation of bolts, nuts and gaskets shall meet the technical requirements for the equipment. At each connection, the portion of high-strength bolts subjected to spot check for torque should not be less than 10 %.

6.1.8 Before the high-strength bolts for structural connection are tightened, the lifting gear shall not be unhooked and shall maintain the necessary holding force. High-strength bolts having accomplished their rated torque values shall not be reused.

6.1.9 The wind turbine installation is not allowed under severe weather conditions such as thick fog, heavy rain or snow, lightning, poor visibility, and low temperature below −20 °C.

6.1.10 The wind speed limit for installing each part of a wind turbine shall meet the technical requirements for the equipment. In the absence of manufacturer requirements, the wind speed limit should be 10 m/s for tower and nacelle installation, and 8 m/s for rotor installation. When the wind speed exceeds the limits, the wind turbine lifting shall be stopped.

6.2 Tower

6.2.1 Before the tower is installed, it shall be checked to make sure that the anticorrosive coating on the surface is intact, the ladder, platform and cable support are installed correctly and firmly, the lighting system is intact, and the tower size deviation meets the design requirements.

6.2.2 The sealant construction at the tower flange shall meet the technical requirements for the equipment, and a layer of sealant should be applied to the outer edges of the flange.

6.2.3 During installation, two adjacent tower segments shall be assembled according to the butt markings on them. The longitudinal welds of two adjacent tower segments should be staggered by 90°.

6.2.4 The verticality of the tower shall be checked after its installation. The verticality of the tower's centerline shall meet the technical requirements for the equipment and comply with the current national standard GB/T 19568, *Wind Turbines— Assembling and Installation Regulation*.

6.2.5 The earthing device shall be installed immediately after the tower connection is completed.

6.2.6 If the nacelle cannot be installed within 12 h after the tower is installed, necessary measures to prevent tower resonance shall be taken.

6.2.7 The installation of the climbing aid and the overhaul lift inside the tower should be conducted after the rotor installation, and shall meet the technical requirements of the equipment.

6.3 Nacelle

6.3.1 The nacelle shall be inspected before installation and meet the following requirements:

1 The nacelle shell is free of any damage, with intact waterproof seal, and clean both inside and outside.

2 The aviation lights, anemometer and lightning protection devices have been mounted on the top of the nacelle.

3 The nacelle lubrication system, yaw system, and shaft brake system are in normal conditions.

6.3.2 When installing the nacelle, a tag line shall be tied respectively at the front and rear of the nacelle, and abrasion-proof measures shall be taken at the contacts between them.

6.3.3 Before installing the nacelle, a trial lifting shall be made to confirm that it is okay and ready for formal lifting.

6.3.4 After the nacelle is lifted off the ground, its levelness shall be adjusted as per the requirements of the manufacturer for further lifting.

6.3.5 Before removing the slings, the torque values of high-strength bolts connecting the nacelle and the tower shall meet the requirements of the manufacturer.

6.3.6 After the nacelle installation is completed, it shall be immediately connected with the lightning protection earthing net.

6.3.7 The generator installation for a direct-drive wind turbine shall meet the following requirements:

1 Inside and outside of the generator shall be kept clean.

2 The generator installation shall use the special lifting spreader provided by the wind turbine manufacturer, and the assembly of the special lifting spreader shall meet the requirements of the manufacturer.

3 Before the generator is installed, the torque values of the high-strength bolts between the tower segments, or between the nacelle and the tower shall meet the requirements of the manufacturer.

4 The tag line shall be fixed at the designated position on the generator to prevent the tag line from being crushed when the generator turns over.

5 The steel wire ropes run through the lifting eyes on both sides of the generator shall be protected to prevent the ropes from wearing the anti-corrosion layer of the lifting eyes.

6 Operators on the generator shall take good safety protection measures and shall not step on the wire protection board of the generator; no people shall stand on the generator when it turns over.

7 After the generator has turned over to take on the angle specified in the technical requirements for the equipment, it can be further lifted for installation.

8 Before removing the lifting spreader, the torques of high-strength bolts between the generator and the nacelle shall meet the requirements of the manufacturer.

6.4 Rotor

6.4.1 When a rotor is stacked on site, the blade length direction should follow the prevailing wind direction, and reliable fixing measures shall be taken.

6.4.2 Before installation, the rotor shall be checked whether the anticorrosive coating on the blade surface is intact, the lightning protection system works normally, and the aviation warning color paint on the blade tip is intact.

6.4.3 The rotor assembly shall meet the following requirements:

1 The hub shall be kept clean, it shall be placed horizontally before assembly, and its pitch system shall work normally.

2 The blade lifting points shall be set at the blade lifting marks. Flat slings should be used, and protective measures shall be taken.

3 The 0° mark on the blade root shall align with the 0° mark on the hub pitch bearing.

4 The blades assembled onto the hub shall be supported using special support members. The supporting points shall be at the support marks. Abrasion-resistant measures shall be taken between supports and blades.

5 The blades shall be assembled according to the blade assembly codes provided by the factory. After the assembly is done, the contact gap between the blade rainproof ring and the hub shell shall meet the rainproof requirements. Sealant shall be applied to each joint of the hub shell.

6 Pitch operation should not be conducted before nuts are installed for the high-strength bolts between the blades and the hub, and torque tightening of the high-strength bolts should be completed on the ground.

7 After a single blade is assembled, the torques of high-strength bolts

between the blade and the hub shall meet the requirements of the manufacturer before removing the slings.

8 If the rotor is unable to be installed immediately after assembly, measures shall be taken to prevent overturning or displacing.

6.4.4 The rotor installation shall meet the following requirements:

1 During installing the rotor, a tag line shall be provided for each blade being lifted by the main hoist, which is fixed at the blade tip with a protective sack, and shall be dragged by sufficient personnel.

2 There shall be no debris left inside the hub before lifting the rotor, and its inside and outside shall be cleaned up.

3 Measures shall be taken to prevent the upper opening angle of the rotor after the auxiliary hoist is unhooked, and the auxiliary hoist shall have protective measures to prevent damage to the blade.

4 When lifting the rotor of a variable pitch turbine, the blades are in the feathered position and locked reliably.

5 When the rotor is lifted about 1.5 m off the ground, the screw holes of the hub shaft flange shall be cleaned.

6 Before removing the slings, the torque of high-strength bolts between the hub and the nacelle or generator shall meet the requirements of the manufacturer.

6.4.5 After the rotor is installed, the brake system shall be adjusted to the working state required by the wind turbine manufacturer. If workers need to enter the hub, the rotor shall be locked reliably before their entry.

6.5 Electrical System

6.5.1 The wind turbine cables and their accessories delivered to the site shall be inspected timely and meet the following requirements:

1 The technical documentation of the products shall be complete.

2 The cable type, specification and length shall meet the requirements of the order, the accessories shall be complete, and the cable appearance shall be intact.

3 The cable terminations shall be closed and tight. To clear up doubts, if any, in visual inspection, an insulation test shall be performed.

6.5.2 When installing the wind turbine, the cable cut shall consider leaving a reserved length near the terminal end and the joint, and the cutoff shall be sealed with a waterproof insulating tape.

6.5.3 The power cables and control cables from the nacelle to the control cabinet at the bottom of the tower shall be installed in accordance with the pertinent requirements of the manufacturer and the current national standard GB 50168, *Standard for Construction and Acceptance of Cable Line Electric Equipment Installation Engineering*. Effective measures shall be taken to prevent cable sway caused by vibration or cable twist when the unit yaws.

6.5.4 Fireproof measures shall be taken at the positions where cables enter and exit the cubicles or holes.

6.5.5 The installation of electrical equipment at the bottom of the tower shall meet the following requirements in addition to the current national standard GB/T 51121, *Code for Construction and Acceptance of Wind Power Project*:

1 Before installing electrical equipment at the bottom of the tower, the levelness of the crane pad shall be checked.

2 The electrical equipment to be installed at the bottom of the tower shall be installed continuously with the bottom section of the tower. If they cannot be installed continuously under special circumstances, measures against wind, rain and dust shall be taken.

6.5.6 The bus duct installation in the tower shall comply with the current national standard GB/T 51121, *Code for Construction and Acceptance of Wind Power Project*.

6.5.7 The electrical wiring and connections shall be firm and reliable.

6.5.8 When assembling a wind turbine, the rotating direction of the generator and the phase sequence at the generator outlet shall be marked, and the phase sequence shall be checked before the initial energization. A cable tag shall be hung at the cable head after it is mounted and bundled up.

6.5.9 The installation of electrical system, protection system, monitoring system, lighting system, and earthing system shall meet the design requirements to ensure that the connections are safe and reliable, and the connection patterns shall not be changed at will.

6.5.10 The installation of other auxiliary electrical circuits shall comply with the design requirements and the manufacturer instructions.

7 Electrical Equipment Installation

7.1 General Requirements

7.1.1 Electrical equipment, installation materials and main structural parts shall be subjected to on-site inspection and unpacking inspection, and shall meet the following requirements:

1 The outer package and seal for the equipment shall be intact.

2 The equipment specifications, models, and quantities shall be consistent with the packing list and meet the design requirements. The accessories and spare parts shall be complete, and the product appearance shall be intact.

3 The technical and quality certificates for products shall be complete and effective.

4 Any inspection shall be kept with inspection records. Any problems found during inspection shall be recorded and solved.

7.1.2 Before the electrical equipment installation, interim handover and acceptance with the civil part shall be completed. The interim handover and acceptance involve the inspection and acceptance of the physical works and related data. For non-conforming items, a list of defects shall be made and handed over to the responsible party for remedy.

7.1.3 Concealed works for electrical equipment installation may be covered up only after passing the acceptance inspection by the supervisor, and the contractor shall record the acceptance inspection for the concealed works.

7.1.4 The earth- or neutral-connection branch lines shall be separately connected to the earth- or neutral-connection trunk line, and shall not be connected in series.

7.2 Prefabricated Substation

7.2.1 The prefabricated substation installation shall comply with the current standards of China GB/T 17467, *High-Voltage/Low-Voltage Prefabricated Substations*; and DL/T 5161.3, *Specification for Construction Quality Checkout and Evaluation of Electric Equipment Installation — Part 3: Power Transformer, Oil Immersed Reactor and Mutual Inductor.*

7.2.2 The connection of prefabricated substation shall comply with the current national standard GB 50149, *Code for Construction and Acceptance of Busbar Installation of Electric Equipment Installation Engineering.*

7.2.3 The earthing mains led from the earthing device shall be directly connected to the neutral on the low-voltage side of the transformer; the earthing mains shall be directly connected to the N and PE buses of the prefabricated substation; the transformer proper and the bracket or enclosure of the dry-type transformer shall be earthed. All connections shall be reliable, and fasteners and anti-loosening parts shall be complete.

7.3 Collection Lines

7.3.1 The installation of overhead lines, electrical equipment on poles, and communication lines shall meet the design requirements and comply with the current national standards GB 50173, *Code for Construction and Acceptance of 66 kV and Under Overhead Electric Power Transmission Line*; and GB 50233, *Code for Construction and Acceptance of 110 kV ~ 750 kV Overhead Transmission Line*.

7.3.2 The laying and wiring of cables shall meet the design requirements and comply with the current national standard GB 50168, *Standard for Construction and Acceptance of Cable Line Electric Equipment Installation Engineerin*.

7.3.3 For tower construction in winter, sufficient safety protection supplies and heating equipment shall be provided, and anti-skid and anti-freezing measures shall be taken during construction.

7.4 Step-up Substation

7.4.1 The installation of primary electrical equipment in the step-up substation shall meet the following requirements:

1 The installation of power transformers, oil-immersed reactors, and instrument transformers shall comply with the current national standard GB 50148, *Code for Construction and Acceptance of Power Transformers Oil Reactor and Mutual Inductor*.

2 The installation of sulfur hexafluoride circuit breakers, gas-insulated metal-enclosed switchgear (GIS), vacuum circuit breakers, high-voltage switchgears, disconnectors and their operating mechanisms, load switches, high-voltage fuses, surge arresters, neutral point discharge gaps, dry reactors, and capacitors shall comply with the current national standard GB 50147, *Code for Construction and Acceptance of High-Voltage Electric Equipment Installation Engineering*.

3 The installation of complete-set reactive power compensation equipment shall meet the design requirements.

4 The installation of 0.4 kV complete-set switch cabinets for the station service system shall comply with the current national standard GB 50171, *Code for Construction and Acceptance of Switchboard Outfit Complete Cubicle and Secondary Circuit Electric Equipment Installation Engineering.*

5 The installation of 35 kV complete-set neutral small-resistance earthing equipment shall comply with the current national standard GB 50147, *Code for Construction and Acceptance of High-Voltage Electric Equipment Installation Engineering.*

6 The fabrication and installation of busbar devices shall comply with the current national standard GB 50149, *Code for Construction and Acceptance of Busbar Installation of Electric Equipment Installation Engineering.*

7 The laying of power cables, fabrication and installation of cable terminals, fire prevention, and flame retardation shall comply with the current national standard GB 50168, *Standard for Construction and Acceptance of Cable Line Electric Equipment Installation Engineering.*

8 The earthing works shall comply with the current national standard GB 50169, *Code for Construction and Acceptance of Grounding Connection Electric Equipment Installation Engineering.*

7.4.2 The installation of secondary electrical equipment in the step-up substation shall meet the following requirements:

1 Control cables, connecting conductors, wiring terminals, etc., shall be adaptable to local environment and operating conditions.

2 The specifications and models of automation components shall be checked and confirmed correct before installation and wiring.

3 The installation of secondary equipment panels and cabinets shall comply with the current national standard GB 50171, *Code for Construction and Acceptance of Switchboard Outfit Complete Cubicle and Secondary Circuit Electric Equipment Installation Engineering.*

4 The installation of relaying protection and security automatic equipment shall comply with the current national standard GB/T 14285, *Technical Code for Relaying Protection and Security Automatic Equipment.*

5 The installation of automatic system dispatching equipment shall meet the manufacturers' specifications and the design requirements.

8 Debugging and Commissioning

8.1 General Requirements

8.1.1 Before debugging and commissioning, a special committee shall be established, and personnel to be involved in the process shall receive professional training.

8.1.2 An outline for debugging and commissioning shall be prepared to specify the contents, requirements and work plan.

8.1.3 Debugging is accomplished singly and jointly. The debugging and commissioning processes shall be recorded in detail.

8.2 Single Debugging

8.2.1 Single debugging shall include the debugging of single electrical equipment and sub-systems in the substation. The debugging items and requirements shall comply with Appendix A of this specification.

8.2.2 A debugging report shall be compiled after completion of the single debugging.

8.3 Joint Debugging and Commissioning

8.3.1 Joint debugging shall include wind turbine debugging, equipment commissioning debugging, integrated automation system debugging, and grid connection system testing.

8.3.2 Wind turbine debugging shall include off-grid debugging, on-grid debugging, and SCADA system debugging. Wind turbine debugging items and requirements shall comply with Appendix B of this specification.

8.3.3 The off-grid debugging of wind turbines shall meet the following conditions:

1 The wind turbine installation inspection is completed and confirmed, and the site is cleared up.

2 The operating environment and meteorological conditions meet the debugging requirements.

3 The earthing devices of the electrical system of the wind turbine shall be connected reliably, and the earth resistance measured shall meet the design requirements.

4 The generator lead-out wires shall be correct in phase sequence, fixed firmly, and connected tightly.

5 Lighting, communications, and safety protection devices shall be complete, all the switches disconnected, and electrical equipment turned off.

6 The ventilation system and cooling system work normally, and the fire-fighting equipment and other associated facilities are complete.

8.3.4 The on-grid debugging of wind turbines shall meet the following conditions:

1 The off-grid debugging is completed and the on-grid debugging requirements are met. The operating environment and meteorological conditions meet the requirements for on-grid debugging.

2 System parameters such as pitching, current changing, and cooling are set according to the requirements of on-grid debugging, and the lock device of the rotor is released.

3 The external system power supply is ready for on-grid debugging.

8.3.5 Before the wind turbines are energized, the phase sequence of the power cables connected inside the tower shall be consistent with that of the prefabricated substation, and the phase sequence colors shall be marked. The insulation between the three-phase cables and the cable-to-earth insulation shall meet the design requirements.

8.3.6 After the wind turbines are energized, the tower internal lighting, lightning protection earthing system, hoisting device, wind measuring device, electrical wiring and electrical setting values, etc., shall be checked and meet the design requirements.

8.3.7 The debugging items and requirements for equipment commissioning shall comply with Appendix C of this specification.

8.3.8 The debugging items and requirements for the integrated automation system shall comply with Appendix D of this specification.

8.3.9 The test items and requirements for the grid connection system shall comply with Appendix E of this specification.

8.3.10 A debugging report shall be prepared by the undertaker after the completion of the joint debugging.

8.3.11 The wind turbines, after the on-grid debugging is completed, shall be subjected to the commissioning test, and the test record shall be made. The commissioning test shall be carried out in three stages: partial trial operation, complete-set trial operation, and trial production.

8.3.12 The commissioning of wind turbines shall meet the following conditions:

1 The installation quality of wind turbines meet the requirements of the manufacturer.

2 The single debugging and joint debugging are completed, all problems and defects are remedied, all parameters meet the relevant requirements, and the wind turbines satisfy the conditions for commissioning.

3 The environmental and meteorological conditions meet the requirements for safe operation.

8.3.13 A large number of wind turbines may be commissioned in batches. Each unit in commissioning shall run continuously for no less than 240 h without failure. If a wind turbine fails to reach full power in this period, a full power test shall be conducted.

9 Quality Inspection and Control

9.1 General Requirements

9.1.1 The quality inspection and control shall cover the whole process of construction, installation, debugging and commissioning of wind power projects.

9.1.2 Before the construction commencement, the project shall be divided into different components according to the project characteristics and submitted to the supervisor for review and approval.

9.1.3 A quality management plan shall be prepared to state the quality objectives, quality standards, quality assurance and control measures, and testing techniques and methods.

9.2 Quality Inspection

9.2.1 The quality inspection of the road works shall comply with the current sector standards JTG F10, *Technical Specification for Construction of Highway Subgrades*; and JTG/T F20, *Technical Guidelines for Construction of Highway Roadbases*.

9.2.2 The construction quality inspection of the civil works shall comply with the current sector standard DL/T 5210.1, *Specification for Construction Quality Acceptance and Evaluation of Electric Power Construct—Part 1: Civil Construction Engineering*.

9.2.3 The installation quality inspection of wind turbines shall comply with the current national standards GB/T 19568, *Wind Turbines—Assembling and Installation Regulation*; GB/T 51121, *Code for Construction and Acceptance of Wind Power Project*; and GB 50168, *Standard for Construction and Acceptance of Cable Line Electric Equipment Installation Engineering*, and relevant specifications of the manufacturer.

9.2.4 The installation quality inspection of electrical equipment shall comply with the current national standards GB 50149, *Code for Construction and Acceptance of Busbar Installation of Electric Equipment Installation Engineering*; GB 50148, *Code for Construction and Acceptance of Power Transformers Oil Reactor and Mutual Inductor*; GB 50147, *Code for Construction and Acceptance of High-Voltage Electric Equipment Installation Engineering*; GB 50171, *Code for Construction and Acceptance of Switchboard Outfit Complete Cubicle and Secondary Circuit Electric Equipment Installation Engineering*; GB 50169, *Code for Construction and Acceptance of Grounding Connection Electric Equipment Installation Engineering*; and GB/T 14285,

Technical Code for Relaying Protection and Security Automatic Equipment.

9.2.5 The installation quality inspection of power collection lines shall comply with the current national standards GB 50173, *Code for Construction and Acceptance of 66 kV and Under Overhead Electric Power Transmission Line*; GB 50233, *Code for Construction and Acceptance of 110 kV~750 kV Overhead Transmission Line*; and GB 50168, *Standard for Construction and Acceptance of Cable Line Electric Equipment Installation Engineering.*

9.3 Quality Control

9.3.1 The construction quality control of a wind power project shall mainly include the quality control at the construction preparation stage and the quality control in the construction process.

9.3.2 The quality control at the construction preparation stage shall include:

1 Establish a sound quality inspection and acceptance system.

2 Establish sound management systems for design change, technical consultation, technical archives, and materials management.

3 Establish a sound inspection system for raw materials, semi-finished products, structural and accessory parts.

4 Prepare special work statements as well as an implementation plan to follow mandatory provisions.

5 Establish a list of quality control points according to the division of the project; propose and explain the prevention and control measures for common quality problems.

6 Ensure reasonable input of resources.

9.3.3 The quality control in the construction process shall include:

1 Strictly implement the technical disclosure requirements. For each work process, technical disclosure shall be conducted first. No construction commencement is allowed without technical disclosure or with ambiguous process and quality requirements.

2 The construction process shall be subjected to the three-level inspection system, i.e. self-inspection, mutual inspection, and handover inspection. The inspection system shall be strictly followed in the work processes. No subsequent process can proceed without inspection of the previous process.

3 For concealed works, the contractor shall notify the supervisor and relevant parties to witness the inspection and acceptance before the concealment.

4 Quality inspection records shall be kept for the construction processes. The records shall be formed along with the physical works, and shall be traceable for key processes.

5 Test results shall be subjected to statistical analysis, and defects if any shall be treated.

6 The semi-finished and finished construction and installation works shall be properly protected.

10 Occupational Health, Safety and Environmental Protection

10.1 General Requirements

10.1.1 A management plan for occupational health, safety, civilized construction, environmental protection, and soil and water conservation shall be formulated according to the actual conditions of the project and environmental characteristics, to set management objectives and propose relative measures.

10.1.2 Construction and management personnel shall receive training on occupational health, safety, civilized construction, environmental protection, and soil and water conservation.

10.1.3 The possible occupational hazards, safety hazards, and environmental hazards shall be identified, analyzed and evaluated before the project starts; preventive measures shall be taken in the construction process; after the construction is completed, the rationality and effectiveness of the preventive measures shall be analyzed and evaluated.

10.1.4 An emergency rescue system shall be established and continuously improved according to risk prevention requirements and project characteristics, to develop an emergency preparedness plan, allocate emergency resources, and propose countermeasures.

10.1.5 In case of a production safety accident or an environmental pollution incident, the accident/incident shall be reported by the procedure, and the rescue, scene preservation, and investigation shall be conducted accordingly.

10.2 Occupational Health

10.2.1 In response to specific site occupational health hazards, reasonable and effective preventive measures shall be taken to upgrade the process technology, improve workplace conditions, provide personal protective equipment, and prevent occupational diseases.

10.2.2 Operators to enter the construction site should take an occupational health examination at a medical institution approved by the health administrative department. Those with occupational contraindications are not allowed to engage in related jobs.

10.2.3 Operators shall be informed of any potential occupational health hazards in the work process they might be exposed to and the consequences, as well as occupational disease prevention measures and treatments.

10.2.4 Effective measures shall be proposed and implemented to control the noise, dust pollution, toxic and hazardous waste, water pollution, etc. that pose occupational hazards.

10.2.5 Good working and living conditions shall be created in the construction site, office area, and living quarters. For workplaces that might cause occupational diseases, warning signs and instructions shall be set up at eye-catching locations.

10.2.6 The occupational health condition of personnel shall be inspected regularly according to the occupational health management plan, and kept on records.

10.3 Safety and Civilized Construction

10.3.1 In view of safety hazards on site, reasonable and effective preventive measures shall be taken to prevent safety accidents; the construction contractor shall formulate applicable safety operating procedures according to the posts, work characteristics, and technical requirements for equipment safety.

10.3.2 Protective measures and warning signs shall be set at dangerous zones such as cliffs, steep ridges, deep pits and high-voltage live lines in the construction site and surrounding areas. Warning signs shall be established at the locations of dangerous installations, hazardous materials, emergency exits, etc. The use of site safety signs shall comply with the current national standard GB 2894, *Safety Signs and Guideline for The Use.*

10.3.3 Site-specific safety and civilized measures and methods shall be prepared according to the construction characteristics of onshore wind farms. The roads, water supply, power supply, temporary construction facilities, and protective measures at the construction site shall meet the requirements for safety and civilized construction. Workers shall be equipped with personal protective equipment. Emergency facilities shall be equipped on site, and their locations shall be marked.

10.3.4 Personnel entering the construction site shall abide by the requirements for safety and civilized construction on site, and must wear protective equipment correctly.

10.3.5 Lifting operations shall comply with the current sector standards JGJ 276, *Technical Code for Safety of Lifting in Construction*; and DL/T 796, *Wind Farm Safety Procedures*.

10.3.6 Electrical equipment shall have reliable earthing measures. Power distribution cubicles and leakage protectors shall be regularly inspected and their status shall be marked, which shall be confirmed before use. Construction power lines shall be wired reasonably, safely and reliably.

10.3.7 The living quarters, office and work areas shall be equipped with fire-fighting facilities and equipment in accordance with relevant national regulations, and set up with fire safety signs and evacuation signs. Fire-fighting equipment and apparatus shall have qualification certificates.

10.3.8 During the construction process, safety inspections including comprehensive inspection, special inspection, seasonal inspection, holiday inspection and daily inspection shall be carried out to troubleshoot and treat potential safety hazards, and kept on records.

10.4 Environmental Protection and Soil and Water Conservation

10.4.1 Environmental protection shall suit the specific conditions of the project and the local environmental characteristics, making good use of temporary and permanent facilities and taking adaptive measures to mitigate the impact of construction on the environment.

10.4.2 Dust control during the construction process shall meet the following requirements:

1 Watering, covering and cleaning measures shall be taken for dust control at the construction site; and the construction roads should be properly hardened.

2 Fine particles such as cement and excavated earth prone to floating in the air, as well as building materials, shall be covered up or stored in an airtight place.

3 Enclosure, dust prevention and reduction measures shall be taken for the concrete batching plant.

10.4.3 The construction liquid waste control shall meet the following requirements:

1 Domestic sewage and wastewater generated in the production process shall be treated according to the discharge standard requirements.

2 Waste oil generated in the production process shall be treated for recycling.

10.4.4 The construction solid waste control shall meet the following requirements:

1 Solid waste generated during the production process shall be classified, stored and treated according to relevant regulations, and shall not be directly buried or incinerated on site.

2 Construction waste and domestic waste shall be cleared off and transported in time, and piled up at designated locations.

10.4.5 The construction noise control shall comply with the current national standard GB 12523, *Emission Standard of Environmental Noise for Boundary of Construction Site.*

10.4.6 In addition to the current national standard GB 50433, *Technical Standard of Soil and Water Conservation for Production and Construction Projects*, soil and water conservation shall also meet the following requirements:

1 The temporary construction facilities shall be reasonably arranged to minimize land use.

2 The construction shall minimize vegetation damage, and restore the landform and vegetation according to the design requirements after the completion of the project.

3 The spoil areas shall not hinder the drainage system on site, and the temporary spoil area shall be covered, retained and enclosed.

10.4.7 Environmental factors shall be checked, monitored and recorded according to the management plan for environmental protection and soil and water conservation.

Appendix A Items and Requirements for Single Debugging

Table A Items and Requirements for Single Debugging

Item		Requirement
Electrical equipment	Power transformer debugging; reactor and suppression coil debugging; current transformer and voltage transformer debugging; circuit breaker debugging; disconnecting switch, load switch and high-voltage fuse debugging; SF_6 GIS debugging; power cable line debugging; overhead transmission line debugging; bushing debugging; insulator debugging; capacitor debugging; insulating oil test; SF_6 gas test; surge arrester test; secondary circuit test; busbar test; earthing device test; low-voltage electrical appliances debugging	In compliance with the current national standard GB 50150, *Electric Equipment Installation Engineering—Standard for Hand-over Test of Electric Equipment*
	Relaying protection and security automatic equipment testing	In compliance with the current national standards GB/T 7261, *Basic Testing Method for Relaying Protection and Security Automatic Equipment*; and GB/T 14285, *Technical Code for Relaying Protection and Security Automatic Equipment*
	Electrical energy metering device testing	In compliance with the current national standard GB/T 1664, *Inspection Regulation of Electric Energy Metering Device On-site Installation*
	Electrical indicator instruments debugging	In compliance with the current sector standard DL/T 1473, *Verification Code of Electrical Indicator Instruments*

Table A *(continued)*

Item		Requirement
Substation sub-systems	Substation service power system debugging	(1) Insulation resistance value shall be tested to be qualified, and relevant indicating meters and computer sampling values shall be checked to be normal; (2) Relevant protection with load is verified satisfactory to requirements; (3) When the system main power circuit is deenergized, the substation service power system can be quickly switched to the standby power supply
	DC system debugging	(1) The insulation resistance value shall be tested qualified and the rectifier shall be checked to work normally; the voltage regulating function, various related instruments and computer sampling values, and the DC bus insulation inspection instruments shall be checked to be normal; (2) When the DC system works normally and the AC power circuit is deenergized, record the DC bus voltage waveform. The DC system can be switched to the energy storage circuit without wave change to supply power normally
	Service lighting system and accident lighting system debugging	(1) The insulation resistance values shall be tested qualified, and the AC bus of the accident inverted lighting system shall be energized, and the power supply circuits shall be checked to work normally;

Table A *(continued)*

Item		Requirement
Substation sub-systems	Service lighting system and accident lighting system debugging	(2) The inverter, related instruments and computer sampling values shall be checked to be normal, and when the AC power supply of the accident lighting system is deenergized, the system can be quickly switched to the DC power supply
	Step-up transformer and high-voltage busbar system energization and debugging	The impact test shall comply with the relevant regulations of the current national standard GB 50150, *Electric Equipment Installation Engineering—Standard for Hand-over Test of Electric Equipment*. The electrical quantities such as voltage and current shall meet the requirements, and the protection functions are normal and without misoperation
	Power collecting system debugging	(1) The earthing and insulation resistance values shall be tested qualified and the line communication shall be tested normal; (2) The impact test shall comply with the current national standard GB 50150, *Electric Equipment Installation Engineering—Standard for Hand-over Test of Electric Equipment*; (3) Electrical quantities such as voltage and current shall meet the requirements, and various protection functions shall be normal and without misoperation

Table A *(continued)*

Item		Requirement
Substation sub-systems	Microprocessor-based fail-safe system debugging	The hardware and software configuration and system functions of the host computer, computer keys, coding locks, communication adapters, printers, etc. shall comply with the current sector standard DL/T 687, *General Specification for Preventing Electric Mal-operation System with Computer* and the grid interconnection requirements

Appendix B Debugging Items and Requirements for Wind Turbine

Table B Debugging Items and Requirements for Wind Turbine

Item		Requirement
Off-grid wind turbine	Basic functions debugging	(1) Hydraulic system: start the hydraulic pump, check the correct action of each component, and adjust the hydraulic pressure of each part to the specified value; (2) Cooling system: manually start each cooling circuit to check their functionality; (3) Lubrication system: manually start the lubrication pump to check the functionality of the lubrication system; (4) Yaw system: in the normal shutdown state of the wind turbine, manually operate the yaw system to yaw clockwise and counterclockwise respectively, observe the yawing stability, and check the working status of the yaw counter; manually yaw the system to meet the triggering conditions of facing the wind and resume automatic yawing, so that the wind turbine shall be able to automatically face the wind; (5) Pitch system: manually pitch and adjust the zero-degree angle of each blade; check the energy storage device of the blade itself to meet the requirement for fast feathering in the case of deenergization from the power grid; (6) Automatic unmooring: manually yaw the wind turbine to meet the triggering conditions of initial unmooring and ultimate unmooring; resume automatic yaw, the wind turbine shall stop and unmoor when the ultimate unmooring condition is triggered; (7) Wind measuring device: adjust the zero-degree position of the wind vane sensor and the zero-degree position of the nacelle; (8) Main control system: The commands sent by the main control system shall be correctly executed by the actuator of the wind turbine; (9) Brake system: artificially trigger normal shutdown and emergency shutdown, and observe the enabling sequence and process of the two brake systems during emergency shutdown, which shall be consistent with the design requirements

Table B *(continued)*

Item		Requirement
Off-grid wind turbine	Safety chain protection function debugging	(1) Emergency stop: When the wind turbine is operating normally, press the emergency stop button, check whether the unit can perform an emergency stop; (2) Vibration module: simulate a vibration signal and make the signal exceed the factory set value, check whether the controller can record and execute the emergency stop command; (3) Over-speed protection: manual operation makes the speed of the wind turbine exceed the speed setting of the over-speed module, and check whether the wind turbine can perform over-speed protection; (4) Cable torsion protection: manually operate the yaw system, when the yawing meets the trigger conditions of the cable torsion protection, and check whether the wind turbine can perform an emergency stop; (5) Grid deenergization protection: simulate power grid deenergization, and check whether the wind turbine can perform emergency shutdown according to the design; (6) Pitch protection: trigger the pitch protection action, and check the feathering condition of the blades and the status of the safety chain
Grid connected wind turbine	Manual grid connection debugging	Switch the converter to debugging mode, manually debug the grid-connected wind turbine according to the debugging manual provided by the converter manufacturer, and check the grid-connected communication function between the main controller and the converter
	Automatic grid connection debugging	Switch the converter to automatic mode, let the wind turbine automatically connect to the grid when the wind conditions are met, and check the automatic grid-connecting capability of the wind turbine
	Power limitation debugging	The wind turbine power shall maintain at the set value after the power is set at a certain value below the rated power

Table B *(continued)*

Item	Requirement
SCADA system	(1) The communication network of the wind power project shall be checked and it shall conform to the design drawings; (2) The SCADA system shall communicate with all the wind turbines normally; (3) The SCADA system shall correctly display the real-time data, historical data and statistical data of the wind turbine; (4) The statement reporting function and charting function shall meet the design requirements; (5) The control functions to start, reset and stop individual wind turbines and the wind power project shall be normal; (6) The active power and reactive power control functions of individual wind turbines and the wind power project shall meet the design requirements; (7) The SCADA system shall communicate with the integrated automation system of the wind power project normally

Appendix C Debugging Items and Requirements for Equipment Commissioning

Table C Debugging Items and Requirements for Equipment Commissioning

Item	Requirement
Primary and secondary circuits phasing and phase confirmation	Primary and secondary circuits shall have distinct phase color identifications, correct phase sequence, polarity and phase positions
Synchronization device testing and debugging	(1) The appearance by visual inspection is normal, the insulation test result is qualified, the analog quantities, input values and displayed values are within the error tolerances, the synchronous switch-closing setting value is verified, and the results of the entire set of generator outlet tests are qualified; (2) Debugging shall comply with the current standards of China GB/T 7261, *Basic Testing Method for Relaying Protection and Security Automatic Equipment*; GB/T 14285, *Technical Code for Relaying Protection and Security Automatic Equipment*; JB/T 3950, *Automatic Accurate Synchronizers*; and DL/T 995, *Testing Regulations on Protection and Stability Control Equipment*
Debugging of the differential protection for step-up transformer and each busbar	(1) Before putting into operation, the circuit wiring, soft and hard pressure plates enabling, and setting values are correct, the values are checked and set correctly, the secondary system current and voltage are displayed normally, and the differential protection outlet action is correct; (2) After putting into operation, test the differential protection current circuit load and the working voltage and current phase sequence, phase identification and phase position, draw a hexagonal vector diagram, measure the U_{ao}, U_{bo}, U_{co} and U_{ab}, U_{bc}, U_{ac} to establish their relationships with I_a, I_b, I_c. The vector phase angle of the same phase current on both sides of the differential protection shall be 180°, and the current circuit shall be correct
Automatic switching of reactive power compensation device	The dynamic response time device shall track and measure the voltage, current, reactive power and power factor of the load in real time, and realize the fast and impact-free switching of the parallel capacitor bank. The dynamic reactive power response time shall not exceed 30 s

Appendix D Debugging Items and Requirements for Integrated Automation

Table D Debugging Items and Requirements for Integrated Automation

Item	Requirement
SCADA system debugging	(1) The quantity, model, and rated parameters of the equipment shall meet the design requirements, and the earthing connections shall be reliable; (2) The functions of remote signaling, remote measurement, remote control, and remote adjustment shall be accurate and reliable; (3) The function of preventing misoperation shall be accurate and reliable; (4) The function of setting value recall, modification and setting group switching shall be correct; (5) The main and standby switching functions shall meet the technical requirements; (6) The communication address shall be correct, and the system shall have good communication and anti-interference ability; (7) The operating status, data and various fault information of the equipment shall be reflected accurately in real time; (8) Equipment with remote start, stop, and active power output adjustment functions shall respond to remote operations in real time, and act accurately and reliably
Debugging of network communication system and dispatching data network	(1) The wind farm communication dispatching data network shall have two routing channels, at least one of which is an optical cable channel; the electrical energy collection device adopts a dedicated power channel; the bandwidth shall meet the design requirements; (2) The optical fiber transmission equipment, PCM terminal equipment, dispatching PABX, data communication network, communication monitoring and other communication equipment of the wind farm directly connected to the power system shall have the same interface and protocol as the system access terminal equipment;

Table D *(continued)*

Item	Requirement
Debugging of network communication system and dispatching data network	(3) Terminal equipment debugging, communication channels, power master station configuration, power system monitoring master station software testing shall meet the grid connection requirements; (4) The telecontrol communication device and the communication management machine are in the same grid-connected dispatch protocol. The communication protocol adopts the power grid standard protocol. Remote measurement, remote signaling, remote control, statistics, time synchronization, setting, event recording, self-test and recovery, communication, electrical quantities collection, microprocessor-based fault recording and other functions shall meet the grid connection requirements; (5) Follow the specifications of IP address interconnection and IP address allocation for each unit; the channel, port, IP address, subnet mask, and gateway are set correctly, the upload rate of real-time and non-real-time data meet the requirements, and the data network is scheduled for encryption and authentication. The four-remotes (remote measurement, remote signaling, remote control and remote dispatching) debugging of the integrated automation background computer monitoring system for wind turbines. The uploading and scheduling of the I/O information point table meets the grid connection requirements
Relaying protection system debugging	(1) Debugging shall comply with the current sector standard DL/T 995, *Testing Regulations on Protection and Stability Control Equipment*; (2) Single debugging of the relaying protection devices shall check the input, output, sampling and other functions of the components, and the setting values shall be verified; the protection actions shall be simulated in the closed state, the switch shall be tripped, and the protection actions shall be accurate and reliable, and the action time shall meet the design requirements; (3) Debugging of the relaying protection devices in groups shall check that the actual relaying protection action logic is consistent with the preset protection logic; (4) The internal and remote communication and interaction functions of the station level relaying protection information management system shall be realized correctly

Table D *(continued)*

Item	Requirement
Power monitoring safety protection system debugging	(1) The safety protection of the power monitoring system shall include physical isolation devices and firewalls at the station level, and shall be able to realize the network security protection functions of the automated system; (2) The operating functions and parameters of the safety protection related equipment of the power monitoring system shall meet the design requirements; (3) The safety protection operation of the power monitoring system shall be consistent with the preset protection strategy; (4) The system debugging shall meet the requirements of the "*Regulations on Security Protection of Power Control and Monitoring Systems*" (Order No. 14 of 2014 by National Development and Reform Commission)
Telecontrol communication system debugging	(1) The power supply of telecontrol communication devices shall be stable and reliable; (2) The signal channel from the station remote control device to the dispatcher remote control device shall be debugged, and shall be stable and reliable; (3) The remote signaling, remote measurement, remote control, and remote dispatching functions of the dispatcher shall be accurate and reliable, and shall meet the grid connection requirements; (4) The main and standby switching functions of the telecontrol system shall meet the technical requirements
Electrical energy information management system debugging	(1) The configuration of the electrical energy acquisition system shall meet the grid connection requirements; (2) The main and auxiliary meters at the power plant gateway shall have the same specifications, models and accuracy, and shall pass the validation by the local electricity metering department; (3) The CT and PT of the power plant gateway meters shall pass the validation by the local electricity metering department; (4) Before the power plant is put into operation, the electricity meter shall be sealed by the local electricity metering department;

Table D *(continued)*

Item	Requirement
Electrical energy information management system debugging	(5) The electrical energy information of the power plant shall be uploaded to the local electricity metering center and the electrical energy information collection system of the power dispatching system in real time and accurately via the dedicated optical fiber channel
Debugging of the whole plant time synchronization system	(1) The system configuration shall ensure that the automatically dispatched equipment and the relaying protection equipment use the same satellite time synchronization system as the power system dispatching department; (2) The clock source, time scale extension unit, and various timing devices for networking the time synchronization system shall satisfy the technical requirements, time synchronization wiring shall be correct and reliable, and time synchronization communication protocols shall be consistent; (3) The master-slave clock source technical index test, event recording function device, microprocessor-based protection device and other timing devices are accurately synchronized in time and meet the design requirements; (4) The resolution of clock synchronization shall meet the requirements of power automation regulations, and the event time reported by the microprocessor-based device shall include information such as year, month, day, hour, minute, second, and millisecond; (5) The master station issues soft time synchronization and manual time synchronization, and the system time of the microprocessor-based device shall meet the requirements of the time synchronization function accordingly
Uninterruptible power supply (UPS) system debugging	(1) The switching function between the main power supply, bypass power supply and DC power supply of the UPS shall be accurate, reliable, and anomaly alarming function is normal; (2) The computer monitoring system shall accurately reflect the operating data and status of the UPS in real time; (3) The dispatching and regulating equipment shall be powered by the UPS and the DC power system in the station. After the AC power supply is deenergized, the operating time of the UPS with load shall be more than 40 min

Table D *(continued)*

Item	Requirement
Sampling value system debugging	(1) The technical parameters and performance of sensors such as the process-level electronic transformers shall meet the requirements of optical sampling values; (2) The bay-level merging unit shall meet the requirements of the substation system voltage, current, and switching values acquisition control, information convergence and provision of standard interface configuration requirements; sampling control, data processing, and communication control shall meet the requirements of the digital sampling system; (3) The station-level system shall have the human-machine interface and remote control system interface; (4) The sampling value system debugging shall meet the technical requirements for a smart substation
Fault recording test	(1) The functions of event recording, networking, remote transmission, receiving station unified clock time synchronization, switching value and analog quantity triggering are normal; (2) The functions of waveform recording, waveform printing, and waveform upload are normal, and there shall be enough channels and be able to record the situations from 10 s before the fault to 60 s after the fault
Wind power prediction system debugging	(1) The short-term wind power prediction and ultra-short wind power prediction functions are normal; the rolling report to the power system dispatching department is normal; (2) The environmental monitoring instrument debugging shall meet the requirements of the product technical specification, the function of the monitoring instrument shall be normal, and the measurement error shall meet the observation requirements; (3) The debugging shall comply with the current national standard GB/T 19963, *Technical Rule for Connecting Wind Farm to Power System*

Table D *(continued)*

Item	Requirement
Security system debugging	(1) The security monitoring system debugging shall comply with the current standards of China GB 50348, *Technical Standard for Security Engineering*; and GA/T 367, *Technical Specification of Video Monitoring Secure System*; (2) The security system debugging with the network communication (by IEC 61850 protocol) as the core, shall complete the collection and monitoring of station-side fence audio and video, environmental data, fire alarm information, SF_6, access control, anti-theft alarm, electronic data, etc., and transmit the above information remotely to the monitoring center or dispatch center

Appendix E Test Items and Requirements for Grid Connection System

Table E Test Items and Requirements for Grid Connection System

Item	Requirement
Active/reactive power control capability test	(1) The power control adjustment range, response speed, and adjustment rate shall meet the current national standard GB/T 19963, *Technical Rule for Connecting Wind Farm to Power System*; (2) The active power control test shall include static debugging, dynamic debugging, whole plant trial operation test, and 24 h whole plant AGC trial operation test; (3) The reactive power adjustment test shall include the static debugging and dynamic debugging
Power quality test	(1) The appearance by visual inspection is normal, the insulation test result is qualified, and the power supply is normal; (2) The error calibration shall include the interharmonics, voltage fluctuation, flicker, frequency, etc., the function calibration shall include the display, record storage, statistics, communications, time synchronization, etc., and the error and function calibration shall meet the current national standards GB/T 14549, *Quality of Electric Energy Supply Harmonics in Public Supply Network*; GB/T 12326, *Power Quality—Voltage Fluctuation and Flicker*; GB/T 15543, *Power Quality—Three-Phase Voltage Unbalance*; GB/T 12325, *Power Quality—Deviation of Supply Voltage*; GB/T 15945, *Power Quality—Frequency Deviation for Power System*; GB/T 18481, *Power Quality—Temporary and Transient Overvoltages*; GB/T 19862, *General Requirements for Monitoring Equipment of Power Quality*; and GB/T 24337, *Power Quality—Interharmonics in Public Supply Network*; (3) The detection method shall meet the current standards of China DL/T 1028, *Verification Code for Power Quality Analyzer*; and GB/T 19862, *General Requirements for Monitoring Equipment of Power Quality*

Table E *(continued)*

Item	Requirement
Low voltage ride-through capability (LVRC) test	(1) The voltage sag caused by grid fault, wind turbine fault ride-through, and wind turbine disconnection from the grid shall meet the basic grid connection requirements; (2) The fault type and voltage level for LVRC test shall meet the grid connection requirements; (3) The active power recovery rate shall meet the grid connection requirements; (4) The dynamic reactive power shall have the ability to inject reactive current to support voltage recovery. The response time of dynamic reactive current control shall meet the grid operation requirements
Voltage and frequency adaptability test	(1) The wind turbine shall operate normally within the voltage and frequency range specified by the power grid; (2) The flicker value, harmonic value, and unbalance of voltage in the normal operation of wind turbines shall meet the current national standards GB/T 12326, *Power Quality—Voltage Fluctuation and Flicker*; GB/T 15543, *Power Quality—Three-Phase Voltage Unbalance*; and GB/T 14549, *Quality of Electric Energy Supply Harmonics in Public Supply Network*; (3) The wind farm operation in different power system frequency ranges shall meet the operation regulations of the specific frequency zone

Explanation of Wording in This Specification

1 Words used for different degrees of strictness are explained as follows in order to mark the differences in executing the requirements in this specification.

1) Words denoting a very strict or mandatory requirement:

"Must" is used for affirmation; "must not" for negation.

2) Words denoting a strict requirement under normal conditions:

"Shall" is used for affirmation; "shall not" for negation.

3) Words denoting a permission of a slight choice or an indication of the most suitable choice when conditions permit:

"Should" is used for affirmation; "should not" for negation.

4) "May" is used to express the option available, sometimes with the conditional permit.

2 "Shall meet the requirements of..." or "shall comply with..." is used in this specification to indicate that it is necessary to comply with the requirements stipulated in other relative standards and codes.

List of Quoted Standards

GB 2894, *Safety Signs and Guideline for The Use*

GB/T 5905, *Cranes—Test Code and Procedures*

GB/T 7261, *Basic Testing Method for Relaying Protection and Security Automatic Equipment*

GB/T 12325, *Power Quality—Deviation of Supply Voltage*

GB/T 12326, *Power Quality—Voltage Fluctuation and Flicker*

GB 12523, *Emission Standard of Environmental Noise for Boundary of Construction Site*

GB/T 14285, *Technical Code for Relaying Protection and Security Automatic Equipment*

GB/T 14549, *Quality of Electric Energy Supply Harmonics in Public Supply Network*

GB/T 15543, *Power Quality—Three-Phase Voltage Unbalance*

GB/T 15945, *Power Quality—Frequency Deviation for Power System*

GB/T 17467, *High-Voltage/Low-Voltage Prefabricated Substations*

GB/T 18481, *Power Quality—Temporary and Transient Overvoltages*

GB/T 19568, *Wind Turbines—Assembling and Installation Regulation*

GB/T 19862, *General Requirements for Monitoring Equipment of Power Quality*

GB/T 19963, *Technical Rule for Connecting Wind Farm to Power System*

GB/T 24337, *Power Quality—Interharmonics in Public Supply Network*

GB 50010, *Code for Design of Concrete Structures*

GB 50026, *Code for Engineering Surveying*

GB 50057, *Code for Design Protection of Structures Against Lightning*

GB 50086, *Technical Code for Engineering of Ground Anchorages and Shotcrete Support*

GB 50147, *Code for Construction and Acceptance of High-Voltage Electric Equipment Installation Engineering*

GB 50148, *Code for Construction and Acceptance of Power Transformers Oil Reactor and Mutual Inductor*

GB 50149, *Code for Construction and Acceptance of Busbar Installation of Electric Equipment Installation Engineering*

GB 50150, *Electric Equipment Installation Engineering—Standard for Hand-over Test of Electric Equipment*

GB 50166, *Code for Installation and Acceptance of Fire Alarm System*

GB 50168, *Standard for Construction and Acceptance of Cable Line Electric Equipment Installation Engineering*

GB 50169, *Code for Construction and Acceptance of Grounding Connection Electric Equipment Installation Engineering*

GB 50171, *Code for Construction and Acceptance of Switchboard Outfit Complete Cubicle and Secondary Circuit Electric Equipment Installation Engineering*

GB 50173, *Code for Construction and Acceptance of 66 kV and Under Overhead Electric Power Transmission Line*

GB 50194, *Code for Safety of Power Supply and Consumption for Construction Site*

GB 50201, *Code for Construction and Acceptance of Earthwork and Blasting Engineering*

GB 50202, *Standard for Acceptance of Construction Quality of Building Foundation*

GB 50203, *Code for Acceptance of Constructional Quality of Masonry Structures*

GB 50204, *Code for Quality Acceptance of Concrete Structure Construction*

GB 50205, *Code for Acceptance of Construction Quality of Steel Structures*

GB 50207, *Code for Acceptance of Construction Quality of Roof*

GB 50208, *Code for Acceptance of Construction Quality of Underground Waterproof*

GB 50209, *Code for Acceptance of Construction Quality of Building Ground*

GB 50212, *Code for Construction of Building Anticorrosive Engineering*

GB/T 50214, *Technical Code for Composite Steel-Form*

GB 50222, *Code for Fire Prevention in Design of Interior Decoration of Buildings*

GB 50224, *Code for Acceptance of Construction Quality of Anticorrosive Engineering of Buildings*

GB 50233, *Code for Construction and Acceptance of 110 kV ~ 750 kV Overhead Transmission Line*

GB 50242, *Code for Acceptance of Construction Quality of Water Supply Drainage and Heating Works*

GB 50243, *Code of Acceptance for Construction Quality of Ventilation and Air Conditioning Works*

GB 50261, *Code for Installation and Commissioning of Sprinkler Systems*

GB 50325, *Code for Indoor Environmental Pollution Control of Civil Building Engineering*

GB 50327, *Code for Construction of Decoration of Housings*

GB 50345, *Technical Code for Roof Engineering*

GB 50348, *Technical Standard for Security Engineering*

GB 50433, *Technical Standard of Soil and Water Conservation for Production and Construction Projects*

GB 50444, *Code for Acceptance and Inspection of Extinguisher Distribution in Buildings*

GB 50496, *Standard for Construction of Mass Concrete*

GB 50666, *Code for Construction of Concrete Structures*

GB 50720, *Technical Code for Fire Safety of Construction Site*

GB 50924, *Code for Construction of Masonry Structures Engineering*

GB/T 51028, *Technical Code for Temperature Measurement and Control of Mass Concrete*

GB/T 51121, *Code for Construction and Acceptance of Wind Power Project*

NB/T 31113, *Code for Construction Organization Design of Onshore Wind Power Projects*

NB/T 31118, *Archives Acceptance Specification for Wind Power Projects*

DL/T 646, *Manufacturing Technical Requirements for Steel Tubular Structures of Substation and Transmission Line*

DL/T 666, Code on Operation of *Wind Farm*

DL/T 687, *General Specification for Preventing Electric Mal-operation System with Computer*

DL/T 796, *Wind Farm Safety Procedures*

DL/T 995, *Testing Regulations on Protection and Stability Control Equipment*

DL/T 1028, *Verification Code for Power Quality Analyzer*

DL/T 1473, *Verification Code of Electrical Indicator Instruments*

DL/T 1664, *Inspection Regulation of Electric Energy Metering Device On-site Installation*

DL/T 5161.3, *Specification for Construction Quality Checkout and Evaluation of Electric Equipment Installation—Part 3: Power Transformer, Oil Immersed Reactor and Mutual Inductor*

DL/T 5161.17, *Specification for Construction Quality Checkout and Evaluation of Electric Equipment Installation—Part 17: Electrical Lighting Device*

DL 5190.1, *The Technical Specification for Electric Power Construction, Part 1: Civil Structure Work*

DL/T 5210.1, *Specification for Construction Quality Acceptance and Evaluation of Electric Power Construct—Part 1: Civil Construction Engineering*

JB/T 3950, *Automatic Accurate Synchronizers*

JGJ/T 10, *Technical Specification for Construction of Concrete Pumping*

JGJ 18, *Specification for Welding and Acceptance of Reinforcing Steel Bars*

JGJ/T 74, *Technical Standard for Large-Area Formwork of Building Construction*

JGJ 79, *Technical Code for Ground Treatment of Buildings*

JGJ 94, *Technical Code for Building Pile Foundations*

JGJ 96, *Technical Specification for Plywood Form with Steel Frame*

JGJ 107, *Technical Specification for Mechanical Splicing of Steel Reinforcing Bars*

JGJ 120, *Technical Specification for Retaining and Protection of*

	Building Foundation Excavations
JGJ 130,	*Technical Code for Safety of Steel Tubular Scaffold with Couplers in Construction*
JGJ 144,	*Technical Specification for External Thermal Insulation on Walls*
JGJ 162,	*Technical Code for Safety of Forms in Construction*
JGJ 276,	*Technical Code for Safety of Lifting in Construction*
JG/T 156,	*Plybamboo Form*
JG/T 163,	*Couplers for Rebar Mechanical Splicing*
JTG F10,	*Technical Specification for Construction of Highway Subgrades*
JTG/T F20,	*Technical Guidelines for Construction of Highway Roadbases*
GA/T 367,	*Technical Specification of Video Monitoring Secure System*